Energy Transfer and Transformation

Reader

Printed in Canada

ISBN: 978-1-68380-518-2

Energy Transfer and Transformation

Table of Contents

Energy Causes Change

Chapter 1

You walk into a dark room. You flip a switch. What changes? The room changes from dark to light.

You are holding a balloon in a quiet place. You stick a pin into the balloon. What changes? The silence in the room changes when you hear a loud sound—*POP*!

A toy car is sitting on top of a table. You give the car a push. What changes? The toy car changes from sitting still to moving.

Whenever there is a **change**, there is some form of **energy** causing the change. Energy is the ability to cause change.

Energy exists in many different forms. It causes different types of changes.

Big Question

Where can we observe evidence of energy causing change?

Vocabulary

change, v. to become different

energy, n. the ability to cause change

Popping a water balloon causes a noise and a splash!

Change Is Evidence of Energy

When you see changes in the world around you, you are seeing evidence, or proof, of energy. Look at the picture. What changes do you think are happening?

As the wave crashes onto the shore, the water moves. That is one change. Imagine you are there on the seashore. You hear the sound of the water splashing against sand and rocks. The sound changes because of the movement of the water and the wind.

Another change is the movement of the sand as it is pushed by the water. And there's a change happening that you can't see—the temperature of the sand is changing as it is warmed by the bright sunlight or cooled by the water.

All these changes are caused by some form of energy.

All of the changes that happen on this seashore are the result of energy.

Energy Occurs in Different Forms

Energy exists in many forms. For example, think of a carousel—a merry-go-round that you might see at a carnival or amusement park.

Motion: The carousel spins around. The horses move riders up and down. All these moving objects have energy of motion.

You can observe evidence of many forms of energy in a spinning carousel.

Sound: As the carousel spins around, it plays loud, cheerful music. The riders laugh and shout to friends. The carousel and the riders are producing sound, another form of energy.

Light: The bright bulbs on the carousel light up the ride. The bulbs give off the form of energy called light.

Heat: If you're riding a carousel, don't touch the bright light bulbs. Why? Because they are hot! The bright bulbs not only produce light, but also give off heat, another form of energy.

Electrical Energy: The carousel, its musical speakers, and its shining lights are all powered by another form of energy. If you look closely around the ride, you will find electrical cords connecting the carousel to a source of electrical energy.

Objects Can Store Energy

When you see the waves crashing on a beach or watch a spinning carousel, you can see changes happening. The changes are evidence of energy. But energy can also be present even when you don't see such evidence.

When a rubber band is not stretched, it isn't causing any change. There is no evidence of energy. But when the rubber band is stretched, it now has the ability to cause changes if it is released, or let go.

The stretched rubber band has **stored energy**. When the rubber band is released, it moves and makes a sound. Those changes are evidence that energy was stored in the stretched rubber band.

Inside a battery are chemicals that store energy. When you put a battery in a flashlight or a toy, the stored chemical energy will be released and cause a change. The energy can make the flashlight light up or the toy move.

> **Vocabulary**
>
> **stored energy, n.** energy that has the ability to cause change at a later time

What happens when one side of a stretched rubber band is released? *SNAP*!

The Stored Energy of Position

Batteries and stretchy materials can store energy because of the materials of which they are made. Any object also can have stored energy because of its position.

Imagine that you pick up an apple and hold it up above your head. Like a stretched rubber band, the apple has the ability to change position. If you release the apple, it will fall.

When an apple hangs from a tree branch or when you hold the apple above the ground, it has stored energy because of its position. The same is true if you hold up a ball or a book or a bucket—each object has stored energy because it has the ability to cause a change once it starts moving.

This apple has stored energy because of its position.

Energy Can Move and Change

Look at the napping cat in the picture. Not very energetic, is he? But there is energy in this scene that you might not notice at first.

The cat sleeps in a patch of sunlight. Light from the sun travels through space and transfers energy all the way to Earth. The sunlight pours through the screen door into the room where the cat is sleeping.

The light enters the room, and when it hits an object, the energy of light is changed to the energy of heat. So, the cat isn't moving, but energy that has moved across millions of miles is now in the room and warming the cat.

In later lessons, you will learn more about how energy can move from one place to another and change from one form to another.

Light and heat are forms of energy that allow this cat to take a cozy nap.

Moving Objects Have Energy

Chapter 2

Everywhere you look, you see things move. Riders pedal bikes. People walk and run. Birds fly. Fish swim along the bottom of the ocean. Rain falls from the sky. Objects move in deep space too. Planets orbit stars. Meteoroids move through space. Another word for the movement of objects is **motion**. Motion occurs when something changes position.

Big Question

How are energy, change, and movement of objects related?

Vocabulary

motion, n. the process of an object changing position

The ball's position changes as it falls through the hoop.

Think About Describing Motion

Moving objects are described by how their position is changing. Scientists describe how fast an object is moving and changing position. Scientists also describe the direction in which an object is moving—left, right, up, down, and so on.

Moving Objects Can Cause Changes

Some objects, such as the basketball players in the picture, cause their own motion by moving their muscles. They jump and run from one end of the court to the other. The movement of muscles causes change.

Other objects, such as the basketball, are put into motion by outside motion. Moving basketball players cause the ball to change position. The player's hand dribbles the ball up and down. Players pass and catch the ball across the court. In all these examples, moving objects cause changes.

The players and the ball in a basketball game are in constant motion, and the motion causes things to change.

Moving Objects Have Energy

When a moving basketball hits a net, *swish*, the net moves. The net changes from sitting still to moving. When a skier slides down a slope of fresh snow, *whoosh*. The snow changes from powdery to hard and packed beneath the skis. When falling raindrops hit the surface of a smooth puddle, *splash*! The surface of the puddle changes from glassy and still to rippling with motion.

Flying basketballs, speeding skiers, falling raindrops, and all other moving objects have energy. How can you tell? You can tell because moving objects can cause change, and energy is the ability to cause a change. All moving objects have **energy of motion**.

Vocabulary

energy of motion, n. the energy an object possesses while it is moving

The motion energy of a falling raindrop changes the surface of a puddle.

People Use the Energy of Motion

Hammers and Nails

Because moving objects can cause a change, a moving object can be useful. People have figured out many ways to use the energy of objects in motion. We cut our food by slicing it with knives, and we cut our grass with spinning lawn mower blades. These objects in motion can cause the changes that result in getting something done.

Imagine trying to push a nail into a block of wood with your hands. That would be nearly impossible to do. Now imagine placing a hammer gently against the head of the nail and pushing the nail into the wood. You still would not move the nail into the wood. But swing the hammer against the head of the nail—*BANG*!

The motion energy of the hammer causes a change when the hammer hits the nail.

The hammer is heavy, and the swing is fast. The moving hammer now has energy of motion. That energy can change the position of the nail, even forcing it into the wood.

Energy from the moving hammer drives the nail into the wood, but what moves the hammer? The motion of the builder's arm provides the energy to change the position of the hammer.

Pendulum Clocks

Engineers, such as clockmakers, have applied the energy of motion for centuries. Clockmakers as early as the 1650s used their knowledge about the energy of a swinging pendulum to help tell time.

Look at the diagram of a clock shown here. The pendulum on the clock swings back and forth, back and forth, *tick-tock*. As the pendulum swings, it causes changes to the position of the clock's gears. The gears, as a result, rotate the hands of the clock. Those hands point to the numbers around the clock's face, allowing us to tell the time.

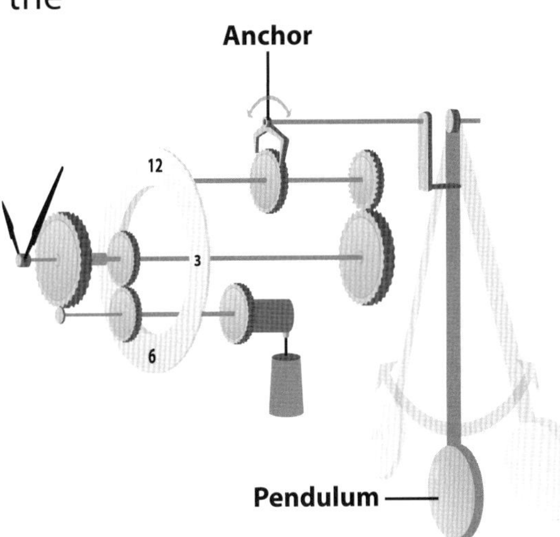

Think About Swinging Motion

The energy of motion of the pendulum arm causes changes in the gears, but what moves the pendulum arm? Gravity! Gravity pulls the arm down as it swings from its highest points on either side. A pendulum clock also has weights or springs that help pull or push on the pendulum.

Wind Turbines

Wind is moving air. Like everything that is in motion, wind has energy. Long ago, people discovered they could use the motion energy from wind to cause changes and get things done. People used wind to push the sails of ships to move across the sea.

Wind has also long been used to push the sails or blades of windmills. Early windmills captured the wind's motion energy to turn the blades. Then, the moving parts inside windmills could, as a result, grind grains or pump water.

Modern wind turbines work in much the same way. Wind turbines use the energy from moving air to spin giant blades. The rotating blades move internal parts that produce electricity, another form of energy.

Both traditional windmills and modern wind turbines use energy from moving air in helpful ways.

Energy and Speed Are Related

Chapter 3

Imagine catching a ball in your hand. If the ball is thrown gently, it doesn't hit your hand very hard. But if the same ball is hurled with strength, it hits your hand hard and stings—*OUCH*! Why does it sting? One reason is because of the greater **speed** at which the ball was moving.

Big Question

How are energy and speed related?

Vocabulary

speed, n. a measurement of the distance an object travels over an amount of time

Speed is a measure of the distance an object travels during an amount of time. One unit for measuring speed is miles per hour (mph). Many countries use kilometers per hour (kph) as a common unit for speed. When you ride in a car traveling 50 mph (about 80 kph), that means you'll move a distance of 50 miles (or about 80 kilometers) in one hour if the car's speed remains constant.

Some pitchers can throw a baseball at speeds faster than 100 miles per hour. Players wear heavy gloves to protect their hands.

Greater Speed Is Related to Greater Energy

Objects with Greater Energy Cause Greater Changes

Remember: energy is the ability to cause change. When you catch a ball, the motion energy from the ball causes a change when it hits your hand. Which will cause greater change—a ball moving slowly or the same ball moving quickly?

The ball moving slowly causes little change—your hand might feel only a gentle tap. Compared to a slow-moving ball, the same ball moving quickly causes greater change when you catch it—your hand moves backward, and you may feel a sting.

Greater changes—the movement of your hand and the stinging sensation—show that an object's speed is related to its energy of motion. The faster the ball is moving, the greater its motion energy. So, if the same ball is thrown twice, the one with the faster speed will have the greater energy. How do you know? You know because there are greater changes caused by the faster-moving ball when you catch it.

The Faster a Ball Rolls, the More Motion Energy It Has

Objects rolling downhill or down a ramp pick up more speed on steeper slopes. Imagine a ball rolling down a ramp. It will have greater energy of motion after rolling down a steep ramp than it will have rolling down a lower ramp of the same length.

How do we know that a fast-moving ball has more energy than that same ball when it is rolling slowly? If we let the ball hit a bucket at the bottom of the ramp, the slower-moving ball from the lower ramp will not move the bucket very far.

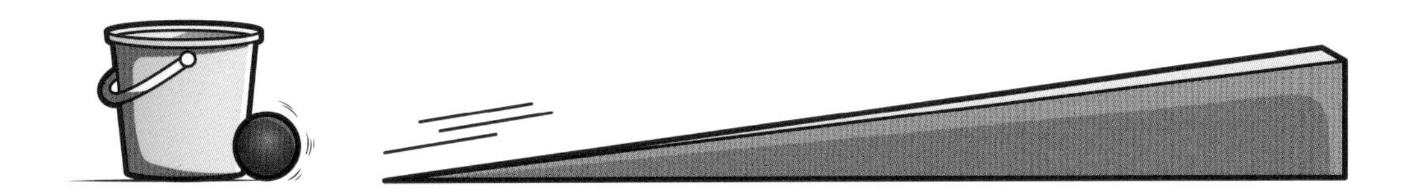

The faster-moving ball from the steeper ramp will knock the bucket over.

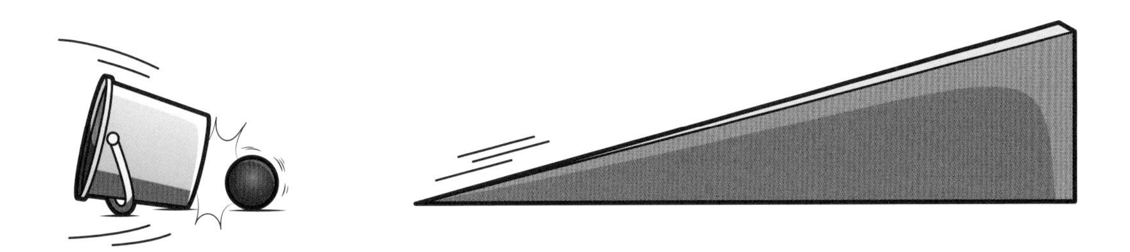

The change to the bucket's position is greater when the ball was moving with more speed. That means the motion energy of the ball was greater when the ball was moving faster.

Bouncing a Ball Shows How Energy and Speed Are Related

You can provide evidence to show that an object causes greater change when it is moving quickly compared to the same object when it is moving slowly.

Bounce a ball. First, gently drop the ball. What changes? The ball bounces, but not very high, maybe up to your knees. And it might make a little noise—*boink*.

Then, throw the ball hard against the ground. What changes, compared to when you gently dropped the ball? The ball bounces higher, maybe as high as your head, and it makes a louder noise—*BOING*!

Now think about throwing the ball against the ground with all your strength, as fast as you can. What changes compared to the other times you threw down the ball? It bounces even higher, way above your head, and it makes an even louder noise—*BWANG*!

The motion energy of a fast-moving ball is greater than when the same ball moves slowly. How do you know? You know because you can compare what changes when the ball hits the ground and bounces. When the ball is moving at a faster speed, you know it has greater energy of motion because you can see and hear greater changes. You see a higher bounce and hear a louder sound. Greater change is evidence of greater energy.

A ball bounced softly will result in less change than one bounced hard.

Energy Transfer

Chapter

4

Your friend is standing still on a skateboard. She asks you to give her a gentle push. Your friend starts out at rest, but when you push on her back she moves down the sidewalk, rolling away from you.

Energy causes that change. When you push your friend, your motion energy causes her to change from being at rest to moving on the skateboard. When you push your friend, you cause an **energy transfer** from you to your friend.

Big Question

What evidence shows that energy transfers from place to place?

Vocabulary

energy transfer, n. movement of energy from one object to another or from one place to another

Word to Know

Evidence is anything that helps prove or disprove an idea. Your own observations or information and facts presented by others can be used as evidence. Scientists and engineers use evidence to support explanations about how and why things happen and to improve their solutions to problems.

Energy Transfers Between Objects in Contact

If you kick a ball so that it hits a second ball, what happens? When you kick the first ball, you transfer energy from your body to the first ball. The position of the ball changes as it moves.

When the first ball hits the second ball, it transfers its motion energy to the second ball. Now, the position of the second ball changes, too. These changes in motion are evidence that energy has transferred from one object to another.

Motion energy transfers from the kicker's foot to the ball, causing a change of the ball's position.

Energy Transfers in Different Ways

All forms of energy—including light, heat, electricity, and sound—are ways that energy transfers, or moves. Energy can transfer from one object to another. Energy can also transfer from one place to another. You have seen examples of energy transfer between objects that touch each other. But are you in contact with the sun when you see its light? Light can transfer energy over long distances through space. This energy transfer happens without the motion or contact of objects.

Light Transfers Energy

Light transfers energy from a light source, such as a light bulb, to your eyes. The energy of light causes changes in your eyes. These changes allow you to see things around you.

The sun is another light source. Light from the sun transfers across 150 million kilometers of space to reach Earth. That's 93 million miles that the sun's light travels to reach us here on Earth! You can see the sunlight reflecting off your skin. Too much of the sunlight can cause an unpleasant change—sunburn! The sunburn is a change, evidence that energy was involved.

Light energy transfers from the sun and causes a change. And a sunburn can be a very unpleasant change.

Sound Transfers Energy

Sound energy often comes from objects that are not touching you, but it causes changes you can hear and sometimes feel. Sometimes when a horn honks nearby or a firework goes off, the sound energy pushes the air, and you feel a push from the sound. Such changes are evidence that an energy transfer happened.

Heat Transfers Energy

Energy also transfers from place to place as heat. If you pour hot tea into a cool cup, soon the cup gets hot. Let a spoon rest in the tea, and soon the spoon gets hot. This is because heat energy transfers from the tea to the cup and to the spoon. The heat energy causes a change. It changes the temperature of the cup and spoon.

The cup and spoon becoming hot is evidence that energy has transferred (moved), because the heat in one material, the tea, moved and produced changes in other objects, the cup and spoon.

The energy of this hot tea transfers through direct contact between the tea, cup, and spoon. Beware of hot food dishes. They can transfer heat to your hands!

Electricity Transfers Electrical Energy

A light bulb without electricity is just a thing made of glass and metal. But when you add electrical energy to the bulb, what changes? It lights up and becomes a source of light, changing a room from dark to brightly lit. The glowing light is evidence that energy has moved from one place to another. The transfer of electrical energy causes the change from an unlit bulb to a lit bulb.

Energy is transferred from place to place by electricity. Like all forms of energy, electrical energy moves from place to place and causes changes. You've probably seen power lines strung between poles along a road. These wires transfer electrical energy from power plants to homes, schools, and businesses. The electrical energy transferred through these wires causes changes when it is used to heat up your toaster or light your room.

Electrical energy transfers through thick wires like these from power plants to the buildings where people use the electricity.

People Transfer Energy to Improve Society

Engineers often solve problems by making energy readily available and useful. Scientists and engineers work to find new and better ways to use different forms of energy and transfer them from place to place. Hydroelectric power plants use a series of energy transfers so that the motion energy of water can be converted to electrical energy.

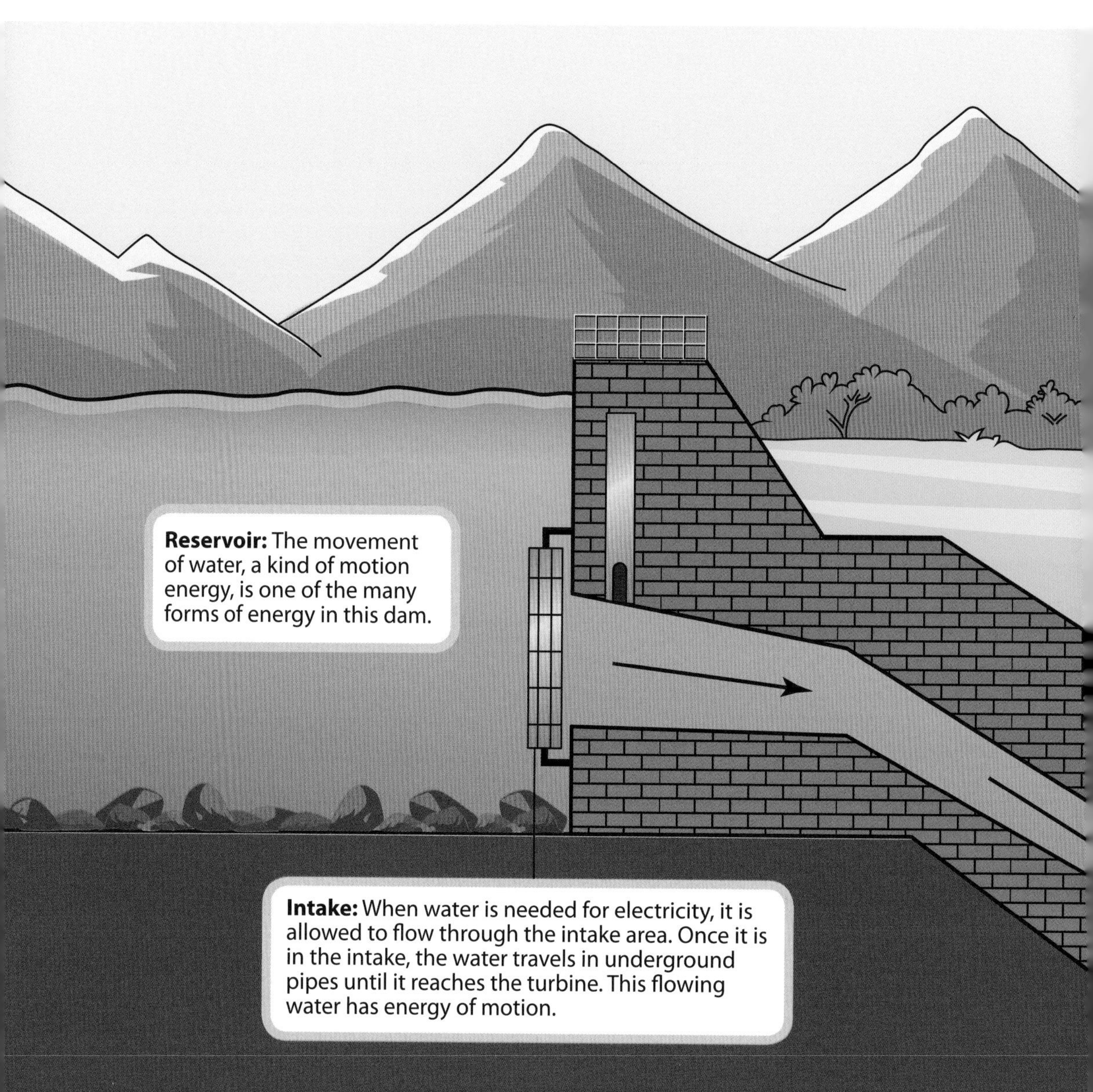

Hydroelectric power uses flowing water to move large, spinning turbines. The motion energy of the turbines causes a change that produces electrical energy. And then the energy of electricity moves through wires over long distances, from place to place.

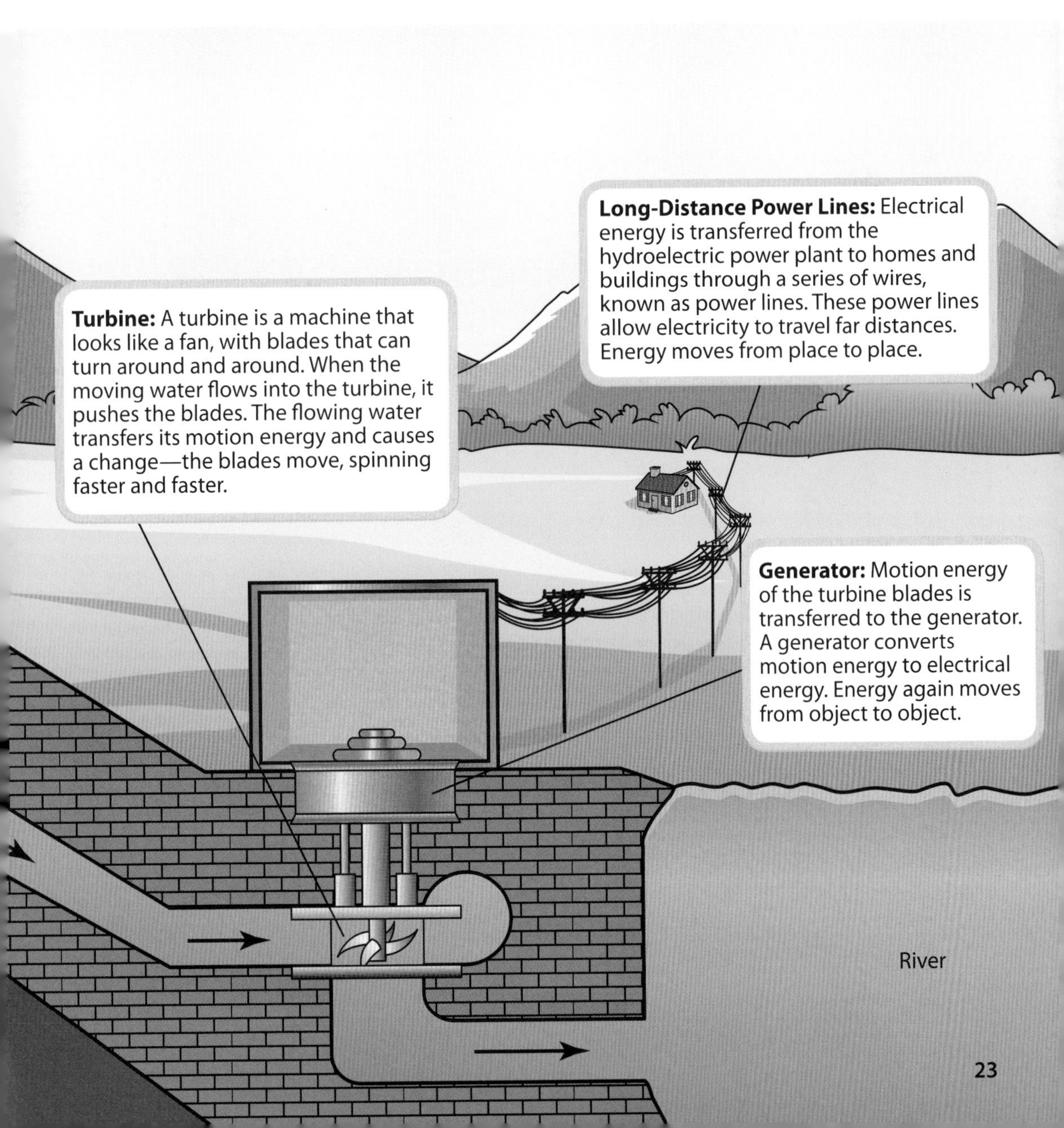

Hoover Dam was built in the Colorado River at the border where Nevada and Arizona meet. You can see in this top photo how the dam holds water back in a great reservoir. When water is allowed to flow through the dam into the lower river, the moving water transfers its energy to equipment inside the dam.

This middle picture of a different dam shows how water moves through a hydroelectric plant with a great deal of energy.

These engineers are building the turbines that go inside a hydroelectric dam. Energy from the moving water transfers to turbines like these. Then the energy transfers in the form of electricity through power lines to buildings where people work and live.

Collisions Transfer Energy

Chapter 5

Clap your hands. When one hand hits the other, your hands **collide**. Lift your foot, and then stomp down to make it collide with the floor. Roll two toy cars down ramps toward each other, and watch them bang into each other. That's a **collision**, too. A collision happens when two objects make contact.

In every collision, changes occur. You might see objects change the speed or direction of motion or even change shape because of a collision. You might hear a sound or see sparks from objects colliding. Or, you might feel an object colliding with you.

Big Question

How is energy involved in collisions?

Vocabulary

collide, v. to come together with impact

collision, n. an instance of colliding

The word part *col-* in *collide* and *collision* means together or with.

When you clap your hands, you can feel the results of a collision. The clap sounds and feels different if you clap slowly or quickly. What other changes occur during hand clapping?

Collisions Involve Changes in Motion

Moving objects have energy of motion—and in any collision, at least one object is moving. We know that energy is involved in collisions because collisions result in changes. Energy causes changes.

A bowling ball has motion energy as it rolls along the lane toward the pins. What happens when the bowling ball crashes into the pins? Can you predict how the pins will change?

The bowling ball's energy of motion is about to cause a change.

Before the collision, the pins are sitting still and not moving. The collision with the bowling ball changes that. The ball strikes the pins, and some of the ball's energy of motion transfers to the pins. This makes the pins gain motion energy and leaves the ball with less motion energy. The collision between ball and pins also produces a loud noise—*CRASH!*—a change in the form of energy from motion to sound.

What would happen if two bowling balls smashed into each other from opposite directions? The colliding balls would bounce off of each other, changing their motion. The speed and direction of each ball would change. The energy of motion would cause changes during this collision.

A Collision Can Change an Object's Shape

An object in motion has energy. In a collision, that energy can cause many kinds of changes, not just changes in motion.

Imagine you have a lump of clay rolled into a ball the same size as a baseball. You throw the clay hard against a brick wall. *SPLAT*! The collision doesn't set the wall in motion. The energy of motion causes a different change. First, the motion of the clay ball stops. Second, the shape of the clay becomes as flat as a pancake!

Think again about the two colliding bowling balls. They are hard enough that they can roll into each other, bounce apart, and have their shapes probably remain unaffected. But think about an egg that is dropped and collides with the floor. The shell shatters, and the liquid inside splatters into a mess. This is a change in shape.

Cars are designed to protect passengers from collisions. The front of the car changes shape more easily than the inside where people sit.

Engineers design cars to protect passengers during a crash. Cars are designed to limit the damage to people that might occur due to a collision. The car in the picture is built so the front part changes shape more easily than the parts closest to the passengers. This design is on purpose. Less of the energy from the collision is transferred to the passengers, better protecting them from injury.

Many Collisions Make Sounds

CLAP! Two hands collide. *CRASH*! Two cars collide. *CRACK*! A baseball and a bat collide. *SPLAT*! An egg and the floor collide. *BONG*! A mallet collides with a drum. Collisions often produce a **sound** as well as changes in motion and shape. When objects collide, some of the motion energy causes vibrations in the objects.

Vocabulary

sound, n. a form of energy that comes from a vibrating object

These vibrations also cause the air to vibrate. Those vibrations transfer energy to our ears. We sense this transfer of energy as sound. When two objects collide, the energy of motion causes changes that result in the energy of sound.

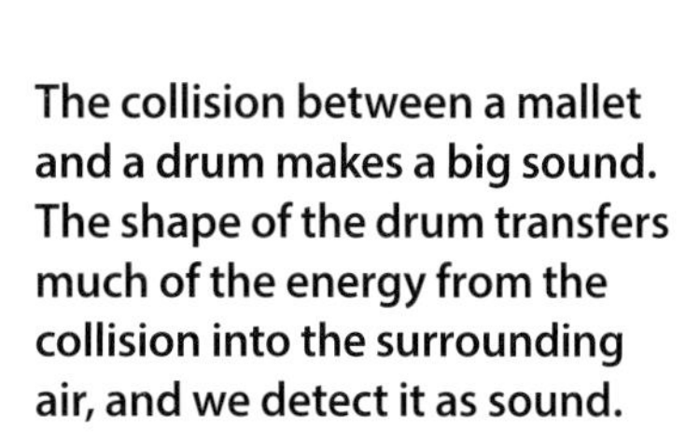

The collision between a mallet and a drum makes a big sound. The shape of the drum transfers much of the energy from the collision into the surrounding air, and we detect it as sound.

Think About Space

Collisions in space don't make sound. There is no air in the vacuum of space. So, there are no particles of matter to be affected by the vibrations of the collision. Without matter surrounding the colliding objects, sound energy cannot move away from the vibrating objects.

Speed Affects Collisions

Speed is how fast an object moves. In previous chapters, you learned that the speed of an object is related to its motion energy. It is not surprising that the speed of moving objects affects the kinds of changes that occur in collisions.

In the game of water balloon toss, the goal is to keep throwing water balloons back and forth without breaking them. The winners know they need to toss the balloons gently and catch them carefully. Throwing a water balloon faster produces a harder collision. Greater energy from the balloon's motion stretches the thin material until it tears and the water gushes out.

An object moving at faster speed has greater energy of motion than the same object moving at slower speed. The object transfers more energy in a collision when it is moving faster than when it moves more slowly. At slower speed, there is less energy of motion that can be transferred.

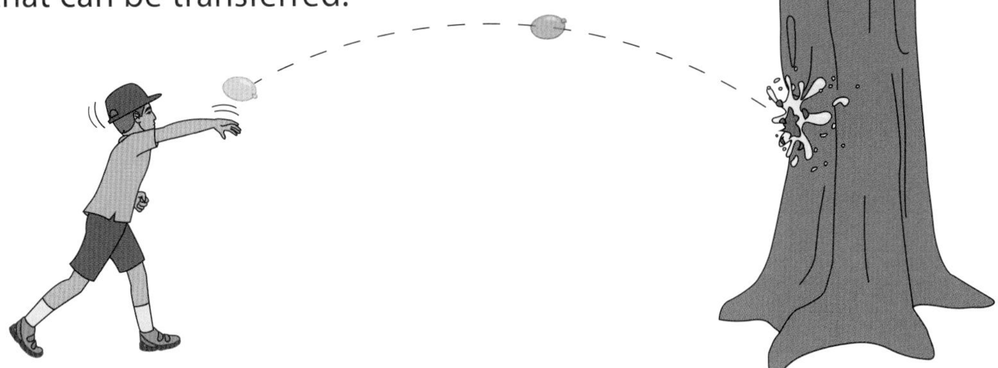

Which picture shows greater speed? Which water balloon had the greater energy of motion?

Weight Affects the Changes Caused by Collisions

Getting hit by a flying cotton ball is silly fun. Getting hit by a flying rock is a different story! Even if the cotton ball and the rock are the same size and moving at the same speed, the rock has greater energy of motion. Why? The rock weighs more. So, the rock has more energy of motion than the cotton ball moving at the same speed. Weight affects the energy of collisions.

Each moon crater offers evidence that motion energy during collisions causes change. Objects in space, such as meteoroids, have been crashing into the moon's surface for millions of years. The heavier these moving space rocks were, the more motion energy they brought to the collisions with the moon. Heavier meteoroids produced deeper craters than lighter meteoroids traveling at the same speed.

The speed and weight of meteoroids have resulted in craters of different sizes on the moon.

How Can Energy Transformations Solve Problems?

Imagine living in a time before toasters. You want to toast a piece of bread. You hold the bread over a fire, but it burns too quickly. Now you have burnt bread! This is a problem. Suppose you need to meet a friend in the morning, but alarm clocks have not been invented yet. You are late, and your friend leaves without you. This is a problem!

Big Question

How do energy transformations help people?

Engineers are people who use science and technology to make things that people want or need. To an engineer, anything that people want or need presents a problem to be solved. All kinds of problems can be solved through the design of new devices. Toasters and alarm clocks are devices that help people. Many such devices are designed to make use of energy transfers and transformations.

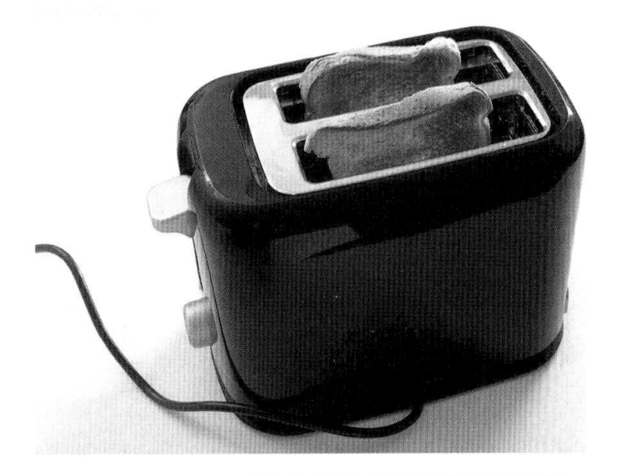

Problem: You want to heat bread to just the right temperature.

Solution: The toaster changes electrical energy to heat, motion, and sound energy. The toasty bread pops up, *DING*, at just the right time.

Solving Problems Requires Understanding Energy

In finding solutions to problems, engineers consider the properties of energy. They know that energy can transfer, or move, from place to place. As a bowling ball collides with bowling pins, some of the ball's energy of motion is transferred to the pins. Your score goes up. Problem solved.

Engineers also know, as do you, that energy can transform, or **convert**, from one form to another. When a bowling ball collides with pins, some of its energy of motion converts to sound energy as the surrounding air vibrates. Such **conversions** occur when energy changes from one form to another.

People use what they know about energy changes to design and build devices that make use of these **energy transformations**. These devices can make life easier, more productive, or even just more fun!

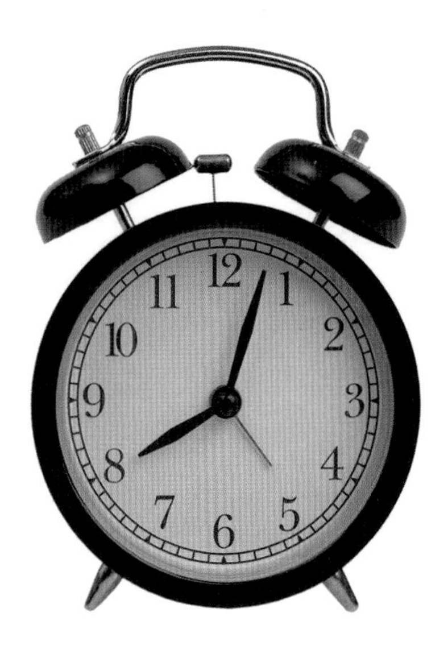

Vocabulary

convert, v. to change form

conversion, n. the process of being converted or changing form

energy transformation, n. the change of one form of energy to another form of energy

Problem: You need to know what time it is.

Solution: Mechanical clocks use springs to store energy. They transform stored energy to motion energy. The moving clock hands show people the time of day.

Transforming Motion Energy Can Help Solve Problems

People need electrical energy to power all sorts of devices, but where does the electrical energy come from? Most of the electricity we use in our homes comes from the transformation of motion energy to electrical energy.

In 1831 a scientist named Michael Faraday discovered that moving magnets past copper wires in certain ways produces a flow of electrical energy through the wires. He discovered a way of transforming motion energy to electrical energy.

Many engineers have used Faraday's discovery to build generators, rotating devices that convert motion to electricity. Hydroelectric dams and the turbines on wind farms use the movement of water and wind to rotate generators. The spinning motion in the generators is converted to electrical energy.

Word to Know

Rotate means to spin. For example, blades of a ceiling fan rotate around and around to make air move in a room.

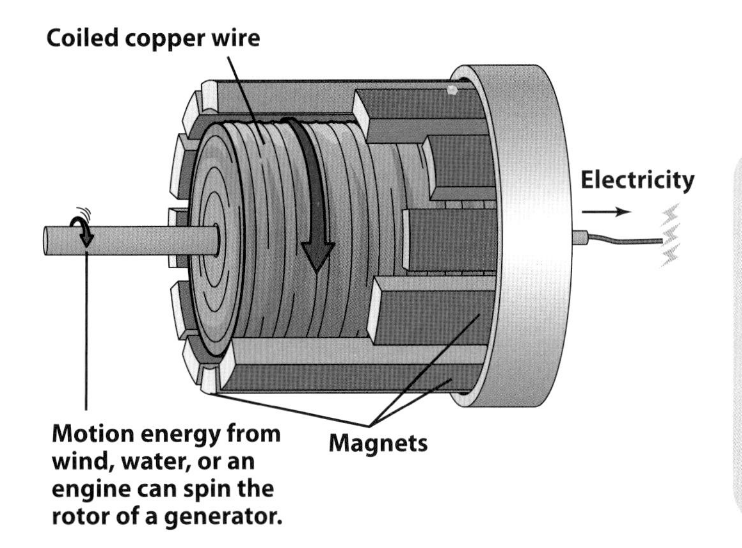

Problem: People need electricity to power many of the devices they use.

Solution: Generators convert motion energy from spinning parts to electrical energy.

Transforming Stored Energy Can Help Solve Problems

Batteries provide electricity, too. Batteries are made of chemicals that have stored energy. The stored energy in a battery transforms to electrical energy. In turn, that electrical energy can be transformed in useful ways by many other devices.

Some batteries "die" after all their stored energy has been transformed to electrical energy. Other batteries can be recharged. Rechargeable batteries can convert electricity from a wall socket to stored chemical energy for later use.

Rechargeable batteries are especially useful for small, portable devices. People want to carry some electronic devices with them wherever they go. These electronic devices use energy transformations in several ways.

A cell phone, for example, converts stored energy in its battery to electrical energy. The electrical energy from the battery transforms to

- sound energy when the phone rings and voices come over the speaker,
- light energy when the screen displays pictures or videos, and
- motion energy when the phone is set to vibrate for notifications.

Problem: People want to be connected to others wherever they go.

Solution: Smartphones with rechargeable batteries store electricity and transform it to sound, light, and motion.

Transforming Light Energy Can Help Solve Problems

Solar cells are made of material that transforms sunlight to electrical energy. The transformation of energy from sunlight to electrical energy also happens on a very large scale. Solar panels convert sunlight to electrical energy during the day. Some of that electrical energy can be stored in batteries to power useful devices at night when the sun is not out.

The solar pad that can charge this phone's battery is small and portable.

A bigger solar panel can convert more sunlight to electrical energy. It can provide electricity for a home that does not have access to electricity generated by a power plant.

Enough solar panels in a sunny region can provide all the electricity needed by an entire community. It doesn't have to be hot to be sunny!

You don't even have to be on Earth for it to be sunny, either! Solar panels solve the problem of how to get the electricity needed to operate the International Space Station.

Engineering Design Solves Problems

All the devices that you have just read about transform energy. They have something else in common, too. They were developed using a process called **engineering design**. Engineering design is used by scientists and engineers to develop solutions to problems.

> **Vocabulary**
>
> **engineering design, n.** a process used to develop a solution to a problem

The process starts with identifying a problem to be solved. Teams of scientists and engineers determine a want or need that the design is going to meet. Then, they judge possible solutions by how practical they are and how well they solve the problem.

For example, electric cars must have rechargeable batteries that store a lot more electrical energy than a phone battery. An electric car battery must store enough energy but not be too heavy or too expensive. Designers have to consider things like cost and weight to solve problems without creating new ones.

Engineers who design electric cars must also design charging stations to transform electricity to stored energy in the cars' batteries.

Solving Problems and Designing Solutions: Thomas A. Edison

Chapter 7

Big Question

How did Thomas Edison use his knowledge of energy transfer and transformation to solve problems?

Thomas Alva Edison changed the world. In his lifetime he designed over one thousand inventions. Edison applied his knowledge of energy transfer and transformation to make an astounding number of useful devices.

Edison knew a lot about the transformation of one form of energy to another. He used this expertise to develop an electric light bulb. His bulb, the carbon filament light bulb, forever changed the way people live.

Edison designed and tested ways to use energy to solve problems. He knew that electrical energy could be transferred through wires. He also knew that, in wires made of some materials, electrical energy would be transformed to light and heat, making the wires glow brightly. Edison wanted to make a light source that would stay lit for a useful amount of time. Edison also wanted his light source to be safe and practical for people to use at home.

Born in Ohio in 1847, Thomas Alva Edison was a busy, curious boy. When he was seven, he left school and was homeschooled by his mother.

Edison Was a Problem Solver

Think about a time before light bulbs. People had to rely on fire to light their way in the dark. Open flames such as torches, candles, and oil lanterns provided the light that people used for any activity after sunset. Edison worked to design a device that could use electricity to be a new kind of light source. Many people didn't want to rely on an open flame that was potentially unsafe. Eventually, after many tests, Edison found a solution.

Edison's carbon filament light bulb—the filament is the wire in a bulb that glows with light when electricity moves through it. The glass protects the filament.

Thomas Edison did not invent the first light bulb. Other scientists and engineers had designed and tested bulbs before Edison. The earliest designs of light bulbs could only provide light for a short amount of time, and the materials were expensive. Edison saw an opportunity to make the early bulbs better and more affordable for homes and businesses. Here was a great need. Thomas Edison improved the designs of others to make his famous bulb.

Edison and his team tested more than six thousand materials to find the right material for light bulb filaments. They wanted to find an inexpensive material that would glow for a long time without burning up. In 1879 he finally found a practical solution: a filament of bamboo with a coating of carbon that transformed electrical energy to light energy.

"When you have exhausted all the possibilities, remember this: you haven't."

Edison thought of engineering designs and then built, tested, and improved what he designed. He often had to conduct many tests before he found a solution that met his goal.

Electrical Energy Where You Need It

Edison's use of energy transfer and transformation extended beyond light bulbs. In 1901 he presented the first alkaline battery. Car batteries at the time used lead chemicals to store energy. Edison wanted to make batteries that would be lighter and more powerful than those heavy lead batteries. He wanted to make electric cars that would be more efficient and reliable. After many experiments and failures, he finally found a combination of chemicals that worked best.

Unfortunately, Edison's idea for electric cars didn't work out at the time. Automobiles designed to burn fossil fuels became more popular. But Edison's work with batteries often still affects your life today. Edison's alkaline battery design was used to develop the batteries you use in flashlights and remote controls.

Materials inside a battery react and transform stored energy to electrical energy. Don't break open a battery! The chemicals are toxic.

These batteries use a reaction between two chemicals to convert stored chemical energy to electricity needed in many devices. Alkaline batteries have long-lasting energy, which makes them dependable. Many modern batteries are the direct descendants of Edison's invention.

Improving Designs Never Ends

Edison famously said, "I haven't failed. I've just found ten thousand ways that won't work." Good inventors, including Thomas Edison, practice persistence. To persist means to keep trying to solve a problem, even when you find ways that won't work. Just as Thomas Edison improved the inventions of others, other inventors built upon his work.

The earliest light bulbs designed by Edison were fragile. Sometimes wires would break off from the filament. Many times, the filament would not last long because of the heat.

Lewis Howard Latimer lived in the same time period as Edison and even worked for Edison at one point. Among Latimer's seven patents are two that improve on the original light bulb design. One of his patents was a design to improve on the way the wire in a bulb was secured to the filament. Another of his patents improved the strength and durability of the filament. This meant the bulb could last longer before the filament broke from the heat.

Lewis Howard Latimer was born in 1848 in Chelsea, Massachusetts. He first worked for Edison's competitors and later for Edison's own company.

Design Solutions over Time

Many kinds of light bulbs waste a lot of the electrical energy that transfers through their filaments. This is because much of the energy changes to heat instead of just light. This is another design problem for engineers to solve. The heat weakens the bulb filament, shortening the time it lasts. The heat also warms the rooms in which people use the bulbs, which isn't welcome in spaces that people try to keep cool.

There are many types of light bulbs today that transform energy a little differently from the way Edison's bulbs worked. Fluorescent and LED light sources use less electricity, produce less heat, and last longer than other bulbs.

Traditional bulbs use electrical energy through a filament to produce light. This produces bright light and a lot of heat.

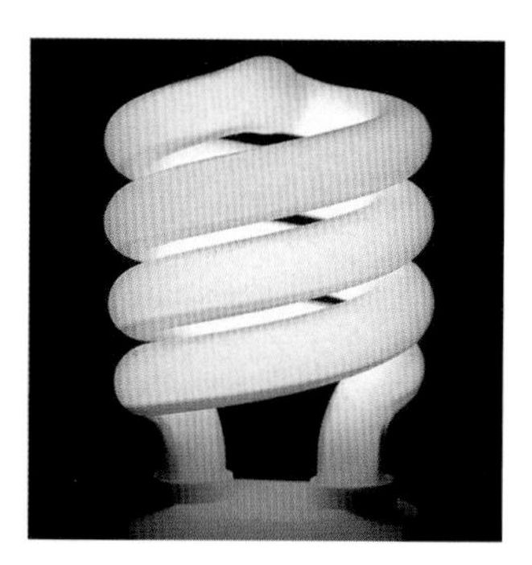

CFL (compact fluorescent light) bulbs use the same technology as neon lights. They give off a glow and little heat instead of bright light.

Tiny LED (light-emitting diode) bulbs are combined in bigger devices. They give off a lot of light and very little heat, using little energy.

Each new design improves an earlier version of the solution—providing light in a better way. In the same way, each version of the battery improves on earlier ways to provide electrical energy. As long as there are problems, people will continue to design solutions.

Glossary

C

change, v. to become different (1)

collide, v. to come together with impact (25)

collision, n. an instance of colliding (25)

conversion, n. the process of being converted or changing form (32)

convert, v. to change form (32)

E

energy, n. the ability to cause change (1)

energy of motion, n. the energy an object possesses while it is moving (9)

energy transfer, n. movement of energy from one object to another or from one place to another (17)

energy transformation, n. the change of one form of energy to another form of energy (32)

engineering design, n. a process used to develop a solution to a problem (36)

M

motion, n. the process of an object changing position (7)

S

sound, n. a form of energy that comes from a vibrating object (28)

speed, n. a measurement of the distance an object travels over an amount of time (13)

stored energy, n. energy that has the ability to cause change at a later time (4)

Core Knowledge®

CKSci™

Core Knowledge SCIENCE™

Series Editor-in-Chief

E.D. Hirsch Jr.

Editorial Directors

Daniel H. Franck and Richard B. Talbot

Subject Matter Expert

Martin Rosenberg, PhD
Teacher of Physics and Computer Science
SAR High School
Riverdale, New York

Illustrations and Photo Credits

Age fotostock / Alamy Stock Photo: 21
Alan Turkus / Flickr: 6
Andrey Armyagov / Alamy Stock Photo: 35
Art of Food / Alamy Stock Photo: 31
B Christopher / Alamy Stock Photo: 42
BC Photo / Alamy Stock Photo: 28
Blue Jean Images / SuperStock: 8
Cultura Creative (RF) / Alamy Stock Photo: 12
Cultura Limited / SuperStock: 24, 35
D. Hurst / Alamy Stock Photo: 32
GODONG / BSIP / SuperStock: 36
H. Mark Weidman Photography / Alamy Stock Photo: 24
Hero Images / SuperStock: 18
Image BROKER / Alamy Stock Photo: 26
Incamerastock / Alamy Stock Photo: 7
Ivy Close Images / Alamy Stock Photo: 39
Juice Images / Alamy Stock Photo: 4
Lubo Ivanko / Alamy Stock Photo: 20
Marjorie Kamys Cotera / Bob Daemmrich Photography / Alamy Stock Photo: 35
Maskot / SuperStock: 35
MPSPhotography / Alamy Stock Photo: 34
National Geographic Image Collection / Alamy Stock Photo: 13
Norbert-Zsolt Suto / Alamy Stock Photo: 42
Oleksandr Kovalchuk / Alamy Stock Photo: Cover B, 5
Paul Mayall / image BROKER / SuperStock: 40
PaulPaladin / Alamy Stock Photo: 9
Pawel Libera / Superstock: i, iii, 3
Peter Ginter / SuperStock: 27
Radius / SuperStock: 19
Russ Bishop / Alamy Stock Photo: Cover D, 2
Science and Society / SuperStock: 38
Science Photo Library / SuperStock: 42
Scott Wilson / Alamy Stock Photo: 24
Stanislaw Pytel / Alamy Stock Photo: 1
Stock Connection / SuperStock: 10
Stocktrek Images / SuperStock: 30
Thomas Baker / Alamy Stock Photo: 11
U.S. National Park Service: Cover A, 37, 41
UpperCut Images / SuperStock: 16